MW00913750

The Future For STEM Is Grim

Why that is true!......How to fix it!

by

George Lopac, Jr.

DORRANCE
PUBLISHING CO
EST. 1920
PITTSBURGH, PENNSYLVANIA 15238

Dorrance Publishing Co
585 Alpha Drive
Pittsburgh, PA 15238
Visit our website at *www.dorrancebookstore.com*

ISBN: 978-1-6386-7033-9
eISBN: 978-1-6386-7982-0

TABLE OF CONTENTS

PREFACE

My first book...*Remaking America Is Up To You – The Future Is Outside The Box*...analyzed why America is a crippled Republic, and presented thirteen paradigm shifts that addressed that situation. This book focuses on a specific element of the U.S. workforce pipeline, STEM skill sets, that is a prerequisite for success in a global economy based on super technology.

The process used to create this book is based on the Common Sense principle: BEFORE YOU DESIGN THE FUTURE YOU MUST LEARN FROM THE PAST AND UNDERSTAND THE PRESENT.

The following pages explain why the future for STEM is grim. Historians present America's evolution from a small group of farmers to a global powerhouse in three major phases of change.

1. Agrarian & Handcraft
2. Industry & Machine Technology
3. Advanced Technology

Each phase was broken down into five segments for analysis.

1. Economic Development
2. Technology Development
3. Educational Development
4. Funding of Education
5. Important Events & Trends

The interactions between those segments, and the effects they have on the future for STEM, are discussed in detail.

SUMMARY

A well-educated Workforce is a prerequisite for a strong economy. The skill sets of a Workforce Pipeline must be in sync with the requirements of a country's developing economy. America is between a rock and a hard place when it must enter a phase of super technology with a Workforce Pipeline that cannot meet the requirements of a super technology economy.

Unfortunately, as America's economy evolved from a small group of farmers to a global powerhouse the focus was on the economy, not the development of a Workforce Pipeline containing skill sets in sync with the economy's requirements. As the economy moved through an advanced technology phase, the need for higher technology skill sets was identified and the STEM movement was born in 2001 (i.e., Science, Technology, Engineering, and Mathematics). After two decades of trying to increase the interest of teenagers in STEM, and teaching teachers how to teach STEM, teenagers' interest in STEM careers is decreasing.

Since the late 1960s the quality of Workforce Pipeline skill sets has been reduced due to abuse of the education system. That is why the future for STEM is grim! Correcting the situation will require serious out-of-the-box thinking to create new paradigms for education, and the monitoring of technology development! Implementation of new paradigms is a necessity, not an option!

INTRODUCTION

Part 1...The Author

It is important that you know that the analysis and conclusions regarding the future for STEM are based on the personal experiences of someone who has experienced all aspects of the STEM story (i.e., student, teacher training, and teacher). It is not the results of academic research done by theorists whose only experience has been with pen and paper. You need to know who I am, what I did, and how I did it! An overview of my career experiences is presented in Appendix A.

During the summer of 1951 I was a twelve-year-old, seventh grade student. My best friend was a thirteen-year-old student scheduled to become a freshman at the Brooklyn Technical High School (BTHS); a public school focused on preparing students for success at college level programs in general College Prep and numerous Engineering curriculums. No one realized it at the time, but BTHS was the first STEM high school in America (i.e., the acronym STEM was not created until 2001). A detailed history of BTHS is presented in Appendix B.

We shared strong interests in airplanes and sports, so it was no surprise that I decided to become a student at BTHS. We both selected the Aeronautical Engineering curriculum as our major interest and graduated in 1956 and 1957 respectively. The education I received was hands down the best in my academic training! Curriculums were rigorous and comprehensive and teachers had experience in the fields they were teaching. My transition from high school to college was trouble-free because of the BTHS experience.

I received my Bachelor's Degree in Mechanical Engineering from the City University of New York...the City College School of Technology...TUITION FREE! I received my Master's Degree from the University of Connecticut. MY FIRST EMPLOYER PAID FOR THE TUITION!

After thirty-eight years of an engineering career I retired and became the teacher of Civil and Mechanical Engineering Technologies for the Middlesex County Academy for Science, Mathematics, and Engineering Technologies; a start-up STEM high school in Edison, New Jersey. Our first year of operation was September 2000 to June 2001.

When the federal government finally realized that the Workforce Pipeline was weak on STEM skill sets, it created the acronym STEM and put out the call for ways to address that problem. The federal government was prepared to fund special programs that addressed the issue. Project Lead The Way (PLTW) was a program designed to teach teachers how to teach STEM. Enticed by possible federal funding, numerous schools sent representatives to PLTW training sessions; a two week period of presentations, workshops, and team exercises. I attended one of the earlier sessions during the summer of 2000. Some problems were immediately identified.

1. The program was designed by engineers for engineers.
2. A large number of participants had no engineering background/training at all.
3. Some participants left the program early because they were frustrated with the mismatch between their experience and the program's content (i.e., why was I sent here?).

I took what I could from the program and went on to develop my own curriculums for grades 9-12 (Ref. Appendix C). Curriculum content was rigorous, comprehensive, and supplemented with personal experiences from my engineering career. After fourteen years of using those curriculums to develop STEM skill sets for the students, feedback from graduates indicated a trouble-free transition to college.

My analysis and conclusions regarding the future for STEM have been based on…

1. My experience as a student at BTHS, a school built for STEM before anyone knew what STEM was.
2. A college education based on rigorous and comprehensive curriculums taught by teachers with experience in the subject they were teaching.
3. Fourteen years teaching rigorous and comprehensive curriculums for civil and mechanical engineering technologies at a STEM high school focused on preparing students for success at the college level.

Part 2…STEM

The acronym STEM was created in 2001, highlighting the mismatch between the skill sets required by the STEM sectors of America's economy and the skill sets of the Workforce Pipeline. It was too little, too late!

For centuries, funding of education has been the responsibility of states and localities. The federal government does not consider education an essential budget item, and teaching is not considered a prime career. The word education did not appear in a federal budget until 1965 as supplementary funds to address concerns with the education process. Primary funding of education has always been the responsibility of states and localities.

The education system suffers from a domino effect. Students moving from grade to grade are not prepared to succeed in the new grade. When high school graduates get into colleges, they are not prepared to succeed in college. That lack of knowledge carries over into the Workforce Pipeline, resulting in the mismatch of skill sets between the pipeline and the STEM sector's future economy requirements.

After two decades of trying to increase teenagers' interest in STEM careers, teenagers' interest in STEM careers is decreasing. There are not enough teachers, teaching STEM subjects, that have professional experience in the subjects they teach. That is not a problem that can be solved by the states and/or localities. It cannot be solved with a Band-Aid! It is a national crisis that requires a total paradigm shift in education (i.e., PreK-12, bachelor's degree, master's degree, doctor's degree, post-doctorate degree, and life-learning). The federal government has to do a better job of preparing for a future economy driven by super technology.

LEARNING FROM THE PAST

The Renaissance

The Renaissance is considered to be the transition from the medieval world to the modern world. It was the great revival of all arts, science, literature, and learning. It started in Italy and continued through the 14th, 15th, and 16th centuries. It was the time when scholars asked the questions; how and why? Knowledge developed during the Renaissance proved to be a catalyst for technology development in America (i.e., 1620-2019).

The Workforce Pipeline

A Workforce Pipeline is a ready pool of workers who are qualified and prepared to step up and fill relevant key roles within the economy as soon as they are identified. A country needs a well-educated Workforce that can respond quickly to changing business needs.

Skills within the Workforce Pipeline are characterized by levels of education.

UNSKILLED…Elementary or no education.
SEMI-SKILLED…High School education.
SKILLED…Vocational/Certification education.
PROFESSIONAL…College education.

A partial list of typical jobs associated with the Workforce skill sets is listed below.

UNSKILLED: Digging ditches…Picking fruit…Mopping floors…Washing dishes… Trash collection… Delivery drivers…House cleaning…all manual labor.

SEMI-SKILLED: Machine operators…Retail sales…Store managers…Cashiers… Home health aides…Typists…etc.

SKILLED: Toolmakers…Chefs…Electricians…Court stenographers…Plumbers… Welders…Mechanics…Technicians…Beauticians…Barbers…etc.

PROFESSIONAL: Teachers…Doctors…Nurses…Lawyers…Engineers…Scientists…
Corporate executives…Statisticians…Accountants…etc.

As we go forward, remember that an economy and its Workforce Pipeline must evolve in sync if the country is to survive the challenges of the future. The problem is defined by two basic questions.

1. How did technology change over time?
2. What events and important trends took place to form the economy?

The Technology Development Curve

In order to analyze America's problem with technical skill sets, it was necessary to have a reliable variable for tracking knowledge development. Plotting the number of patents granted in a given year versus time resulted in an increasing trend with time. But there seemed to be something more taking place. Plotting the number of patents granted in each decade (i.e., a ten-year total) revealed an array of four distinct straight lines and proved to be the reliable variable needed. The number of patents granted in each decade was used to develop the Technology Development Curve (Ref. page 3).

The Technology Development Curve covers the period from 1790 (the start of recording patent grants) through 2019 (Ref. Appendix D).

U.S. TECHNOLOGY DEVELOPMENT

Total Patents vs Decades

NOTES

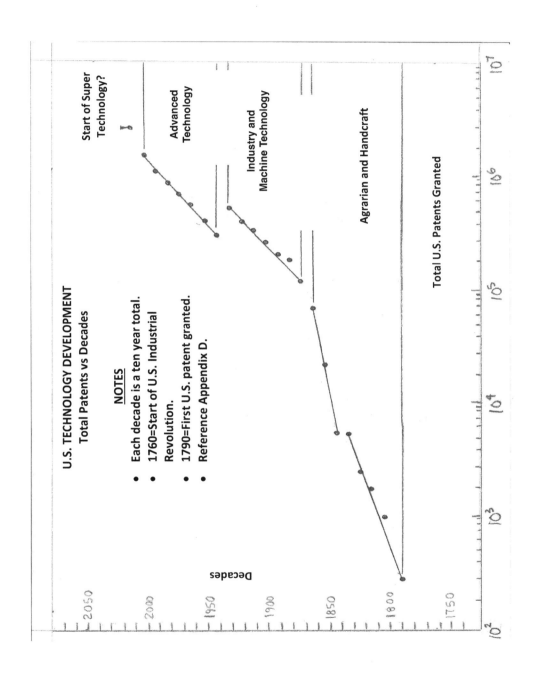

U.S. TECHNOLOGY DEVELOPMENT
Total Patents vs Decades

<u>NOTES</u>

- Each decade is a ten year total.
- 1760=Start of U.S. Industrial Revolution.
- 1790=First U.S. patent granted.
- Reference Appendix D.

Start of Super Technology?

Advanced Technology

Industry and Machine Technology

Agrarian and Handcraft

Total U.S. Patents Granted

Decades

- Technology Development was dormant from 1620-1729.
- Technology Development was in the nascent stage from 1730-1789.
- The start of the American Industrial Revolution in 1760 was the catalyst for continuous, increasing, Technology Development for the balance of the Agrarian and Handcraft phase.
- The completion of the Transcontinental Railway was the start of a major change in the rate and magnitude of Technology Development, signaling the start of the Industry & Machine Technology phase.

The above observations resulted in clarification of the dates for the historical phases of Technology Development, and revealed the starting point of the next phase of Technology Development; Phase 4…Super Technology.

The total number of patents granted from 1790-2019 was 11,386,544.

Phase 1…Agrarian & Handcraft…1620-1869 . . . . .112, 294

Phase 2…Industry & Machine Technology…1870-1939 2,199,741

Phase 3…Advanced Technology…1940-2009 5,941,312

Phase 4…Super Technology…the 1st decade…2010-2019 3,133,197

The total number of patents granted from 2010-2019 was 28% of the total from 1790-2019, and 53% of the previous seven decades, 1940-2009.

A strong case can be made for making the start of Phase 4…Super Technology…The decade from 2010-2019.

A more detailed discussion of these results is presented in the sections to follow.

Phase 1…Agrarian & Handcraft
1620-1869
ECONOMIC DEVELOPMENT

- Started when the Pilgrims landed at Plymouth Rock in 1620.
- As settlers moved across the country, villages became towns, and towns became cities.
- In the villages, bartering was the economic model: to trade by exchanging goods and services without using money.

- Communication was face to face.
- When villages, towns, and cities came into play, a need to communicate with each other developed.
- It also created a need to trade goods and services with an exchange of money, and to transport materials and products between villages, towns, cities.

TECHNOLOGY DEVELOPMENT

- America was pre-industrial throughout the first third of the 19th century. Most people lived on farms and produced much of what they consumed. A considerable percentage of the non-farm population was engaged in handling goods for export. The country was an exporter of agricultural products.
- Technology Development was dormant from 1620-1729.
- Technology Development was in the nascent stage from 1730-1789.
- The start of the American Industrial Revolution in 1760 was the catalyst for continuous, increasing, Technology Development for the balance of the Agrarian & Handcraft phase.
- Technology Development in the 1830s and 1840s was essentially the same due to a number of disrupting events.
  - The 1843 Oregon Fever…the "Land Rush" for a better life out west.
  - The 1846-1847 war with Mexico.
  - The 1848 "Gold Rush".
- That period of constant development resulted in the start of a second straight line in Phase 1 on the Technology Development Curve.
- The Civil War (1861-1865) was fought with a low level of technical knowledge.
- America's climb to industrial power started in 1800.
- America built the best ships in the world.
- The textile industry became established in New England where there was abundant water power.
- Steam power began being used in factories, but water was the dominant source of industrial power.
- The building of roads and canals, the introduction of steam boats, and the first railroads, were the beginning of a transportation revolution that would accelerate throughout the century.

EDUCATIONAL DEVELOPMENT

- During the early stages of Phase 1, education took the form of home-schooling, single-room schoolhouses, and apprenticeships with local craftsmen. The focus was on the basics (i.e., reading, writing, and arithmetic), and learning a trade.

- Teaching the basics was considered woman's work, and teaching an apprentice a trade was highly regarded.

- As the population grew the home-schooling and one-room schoolhouses became obsolete, giving way to larger schools and students being placed in classes with children of the same age. Vocational training still went the way of apprenticeships.

- The first Board of Education was established in 1647.

- State colleges were established. Baruch College in New York was founded in 1847 as the Free Academy; the first free public institution of higher education, according to the college.

FUNDING OF EDUCATION

- Throughout Phase 1 funding education was the responsibility of the states and localities.

IMPORTANT EVENTS AND TRENDS

- The Lewis and Clark expedition reached the Pacific Ocean in 1805, opening the west to the movement of goods and services and increased Technology Development.

- The Colonial and Indian Wars established a war culture that exists today (Ref. Appendix E).

- A technical component of the Workforce Pipeline was starting to form as graduates from new technical colleges entered the workforce.

- The Civil War provided the North with an opportunity to establish and dominate America's industrial and economic future.

- The Northern states' economy quickly proved to be more effective at supporting and sustaining a war economy. Nearly every sector of the Union economy increased production during the Civil War years including, to the detriment of the South, the mechanization of farming.

- America's population grew from 2,302 in 1620, to 38,558,371 in 1870 (Ref. Appendix F).

Phase 2...Industry & Machine Technology
1870-1939
ECONOMIC DEVELOPMENT

- Growth of the economy progressed with regard to agricultural and manufactured production, international trade, federal debt, taxation, transportation, education, and land expansion.

- The government controlled the economy during WWI by allocating scarce materials, coordinating purchasing, determining priorities, encouraging the development of new facilities, and fixing prices.

- When WWI began, the U.S. economy was in recession. A forty-four-month economic boom ensued from 1914-1918; first as Europeans began purchasing U.S. goods for the war and later as the U.S. itself joined the battle.

- Between 1870 and 1912, a period of forty-two years, industrial production in the United States rose by 682%.

TECHNOLOGY DEVELOPMENT

- In 1870 the completion of the Trans Continental Railway opened up America's west for development.

- An explosion of new discoveries and inventions took place.

- The plastics industry was born.

- Numerous plastic materials were invented or discovered, and the search for processes to fabricate and assemble plastic parts was started. The age of Plastics was off and running.

- The assembly line was invented by Henry Ford.

- The concept of Quality Control was implemented.

- America's industrial power continued to grow at a fast pace.

EDUCATIONAL DEVELOPMENT

- Vocational schools were started in 1889 to address the growing need for a Workforce with technical skills.

- Brooklyn Technical High School (BTHS) was established in 1922. It was the first STEM high school, but no one knew it at the time. The history of BTHS is presented in Appendix B.

- Chemical, Mechanical, and Electrical Engineering college curriculums were developed, and colleges started to contribute to the technical component of the Workforce Pipeline skill sets.

FUNDING OF EDUCATION

- Was still the responsibility of states and localities.

IMPORTANT EVENTS AND TRENDS

- The war culture continued to increase its impact on the American economy (Ref. Appendix E).
- WWI...technology mechanized death. The use of airplanes, machine guns, and other advances in military weapons represented a paradigm shift in military operations.
- The technical component of the Workforce Pipeline was developing as a result of vocational training and college level engineering programs.
- America's population had grown from 38,558,371 in 1870 to132,164,569 in 1940.

Phase 3...Advanced Technology
1940-2009
ECONOMIC DEVELOPMENT

- Wars continued to have an effect on the economy (Ref. Appendix E). WWII highlighted the need for optimum coordination between military and industrial operations, initiating the evolution of the Military-Industrial-Complex.
- In the 1940s the economy overall grew by 37%. At the end of the decade the median American family had 30% more purchasing power than at the beginning. Inflation was minimal, and unemployment remained low, about 4.5%. Many factors came together to produce the '50s boom.
- In a post-WWII economy, the United States experienced phenomenal economic growth. The United States consolidated its position as the world's richest country. More and more Americans considered themselves part of the middle class. America had completed its transformation from agricultural to consumerism!
- The GI Bill gave veterans an affordable college education, providing a pool of highly-educated employees for the Workforce Pipeline.
- Cheap oil from the United States' wells fueled industry.
- Advances in science and technology improved productivity while competitors in Europe and Asia were still recovering from WWII.

- By the 1960s the post-war boom had flourished for over a decade and had begun to wane. Under conservative Eisenhower the nation had grown, but only cautiously.

- When Kennedy swept into office his energy and enthusiasm inspired Americans to take on challenges both foreign and domestic. Kennedy's goals were to stimulate the economy, reduce unemployment, support growth and democracy abroad and establish an important economic position on the international front.

- Later Johnson added the goals of eradicating poverty, integrating women and minorities into the economy, and winning the war in Vietnam.

- Environmental and consumer interests were increasingly taken into consideration.

- As the nation strove to achieve these goals, the economy suffered from their negative effects. Large-scale government spending and the constraints of the international monetary system resulted in domestic inflation. As the government struggled to slow inflation and stabilize the economy, the Vietnam War and the war on poverty raged on. One war was hopelessly lost, and the other was only partially won. Optimism disintegrated as the dollar lost stability and inflation took a firm hold.

- By the end of the 1960s, the economy was very different from its state at the beginning of the decade. Growth was slowing, inflation was rising, and the dollar was in poor shape. Nevertheless, there were positive changes in the economy. The United States had obtained greater access to trade with foreign nations. Developments in computers contributed to the increasingly widespread use of computerized technology in business. Women and minorities were increasingly a part of important economic activities. Poverty had been seriously reduced. Legislation to protect consumers and the environment from unsafe business practices was established. The American economy had been thrust into the second half of the century, not fully capable of meeting its challenges, but willing to try.

- In the sixty years after WWII, the United States built the world's greatest middle-class economy; then unbuilt it. Average family income changed at each rung of the economic ladder from 1950 through 2010. In the immediate post-war period, America's rapid growth favored the middle classes. The poorest fifth of all households, in fact, fared the best. Then, in the 1970s, amid two oil crises and awful inflation, things ground to a halt. We cut taxes. Technology and competition from abroad started whittling away at blue collar jobs and pay. The financial markets took off. And so when growth returned, it favored the investment class, the top 20 percent, especially the top5 percent, and the top 1 percent more than anybody. And it all fell apart. 2000-2009 was a lost decade for families, and it's not clear how much better they will fare in 2010.

- As more baby boomers live longer, the service industry demand on health care is expected to be the fastest growing segment of the United States' economy. This evolution in industry growth will be worth tracking as service based sectors gain more of the lion's share of the economy leading into 2010.

TECHNOLOGY DEVELOPMENT

- The number of patents granted in the 1940s was less than the number granted in the 1930s. That anomaly could have been the result of a number of disrupting events.
  - WWII
  - The Atom Bomb
  - The Roswell UFO incident
- The federal government was late in responding to the 1950s Technology Boom; the transition from analog to digital technology.
- The 1950s Technology Boom highlighted the need for more STEM workers in the Workforce Pipeline.
- As time moved on the demand for technology kept growing. All the latest and greatest things were coming out and everyone wanted the newest technology. A partial list of the products and technologies developed during the period from 1940-2009 is presented below.
  - Microwave Technology
  - Radar Technology
  - First computer made
  - Play-Doh in a can
  - Television makes its debut
  - The Manhattan Project started the nuclear age
  - The TV remote was invented
  - First cardiac pacemaker was created
  - The Boeing 707-120 was invented
  - Black and Decker created the first cordless drill
  - The communication Satellite was invented
  - Unarmed Aerial Vehicles were invented

- The smoke detector was invented
- The internet was put to use
- Digital music was created
- Electronic ignition was invented
- MRI was invented
- The PC was created
- The first Laptop was invented
- Atari 2600 was invented
- GPS was created
- The CD was invented
- DNA fingerprinting was invented
- The MP3 player was invented
- The iPod was created
- Bluetooth headsets were invented
- HDTVs were invented
- Blue-ray technology was invented
- Nintendo Wii was invented
- UAB flash drives were invented
- Blackberry technology was developed
- The Kindle e-reader was invented
- Microsoft Windows XP was developed
- Apple Mac OS X was developed
- The iPhone was invented
- Netbooks were invented
- 3D Cinema was developed
- Inductive power gets real
- Operating systems bulk up, then slim down
- App stores are all the rage
- Android OS appears in everything

- The growth of America's industrial power continues.
- America's growth in Automated Manufacturing started in the 1950s and continues.

EDUCATIONAL DEVELOPMENT

- K-12 schools became prominent.
- Vocational schools continued to deliver technical skill sets to the Workforce Pipeline.
- Colleges with Chemical, Mechanical, Electrical, etc., engineering curriculums continued to contribute to the Workforce Pipeline skill sets.
- Up until the late 1960s state universities were tuition-free.
- Changes began after WWII as the GI Bill increased the number of Americans wanting to go to college, and continued into the 1960s culminating in Civil Rights and student protests. These events, the new influx of college eligible Americans, and their demand for education (which outpaced supply and funding), led to the end of tuition-free state universities and the start of a pay for education business model for state universities; and the start of the student loan crisis.
- Before the for-profit business model, requirements for admission to, and remaining in, college were rigorous and comprehensive; resulting in graduates with significant skill sets that were in sync with the Workforce Pipeline requirements.
- The conversion from tuition-free education to a for-profit business model resulted in lowered requirements for college admission (i.e., empty seats do not generate cash flow), lowered requirements for staying in college(i.e., empty seats do not generate cash flow), and reduced quality of the skill sets of graduates entering the Workforce Pipeline. America's talent pool was being filtered based on the ability to pay.
- Since the late 1960s the quality of Workforce Pipeline skill sets has been declining as poorly trained college graduates entered the Workforce Pipeline while "old school" college graduates (i.e., rigorous and comprehensive curriculums) were leaving the Workforce Pipeline (i.e., retirement, down-sizing, etc.). By the early 2000s,the Workforce Pipeline contained 100% of poorly trained college graduates with questionable skill sets.

FUNDING OF EDUCATION

- Funding was still the responsibility of the states or localities. It wasn't until the mid-1980s that the federal government took a more robust role in K-12 education, responding to constant reports of poor student performance since the 1960s. The government funding was for special projects focused on solving specific problems. The primary funding of education was still the responsibility of the states or localities.

IMPORTANT EVENTS AND TRENDS

- Wars continued to drain financial resources and abuse the human resources of the United States.

- The federal government's indifferent attitude toward the educational process and its importance to development of a strong economy is a critical problem.

- The relative decline of American education has long been a national embarrassment, as well as a threat to the nation's future. America's students are not progressing to catch up to their peers in other industrialized countries.

- In 2001 the acronym STEM (i.e., Science-Technology-Engineering-Mathematics) was created in reaction to a technology boom in 2000.

- The federal government put out the call for help in stimulating the technical component of the Workforce Pipeline. The promise of federal funding for special projects got the attention of consultants and Project Lead The Way (PLTW) was created. The goals of PLTW were to provide transformative learning experiences for PreK-12 students and teachers across the United States, create an engaging hands-on classroom environment, empower students to develop in-demand knowledge and skills they need to thrive, provide teachers with the training, resources, and support they need to engage students in real-world learning.

- After two decades of trying to increase the interest of teenagers in STEM, and teaching teachers how to teach STEM, teenagers' interest in STEM careers is decreasing and there is a shortage of teachers that can teach engineering.

- America's population had grown from132,164,569 in 1940 to 308,745,538 in 2010.

Phase 4...Super Technology
2010-2019
ECONOMIC DEVELOPMENT
2010

- 2010 was a year of only slight progress for the unemployed but real gains for workers who still had jobs.

- Without a surge in demand for goods and services produced in the United States, it is hard to see much improvement in the outlook for the unemployed.

- The pace of economic growth must increase before we can see a sizeable drop in unemployment.

- The growing gap between the fortunes of those with and without jobs is matched by the widening divide between Americans who work for others and the businesses that employ them.
- Measured in nominal dollars, corporate profits now exceed their pre-recession levels.
- Business profitability has returned, but total wage and salary disbursements remain lower than they were before the recession began.
- The political danger facing the unemployed is that surging business profitability and the improving fortunes of employees will cause voters to lose sight of the daunting problems confronting the long-term unemployed.

2011

- 2011 for the United States economy was a year of slow growth and fears of a double-dip recession, but there were some more positive signs as 2011 came to a close.
- The debt ceiling conflict was resolved, and it was agreed that government spending would be cut by hundreds of billions of dollars over the next decade.
- American Jobs Act…an ambitious plan to boost employment in the United States economy was introduced by President Obama.
- Operation Twist…a stimulus measure designed to drive long-term interest rates down, thereby encouraging more mortgages and business loans to be made.

2012

- In 2012, business leaders waited for the outcome of one uncertainty after another. As a result, it seemed the economic spark never got oxygen to really burst into flame, even though the fuel was there.
- The biggest contributor was the 2012 presidential election. It was a very close race between two candidates with radically different approaches to stimulating economic growth. The race itself slowed growth as businesses waited to see what direction the country would take.

2013

- The first official estimate of United States growth in 2013 indicated that real GDP, all the economic value produced in the country, increased by just 1.9%.For the optimists the fourth quarter saw the economy growing at a clip that would increase GDP 3.2% if it continued for a full year.

- Personal expenditures were the single largest contributor to United States GDP growth. This was mostly purchasing more goods, with spending on vehicles playing a big role, services growth was important as well, with health care unsurprisingly taking a top spot.

- Private investment came in next. That included investment in homes, business equipment, and computers.

- Inventories did not make a big contribution. Private inventories contributed 0.19 percentage points to growth on the year, about the same as 2012.

- Trade continued to boost GDP. United States net exports contributed 0.14 percentage points of new growth, thanks mostly to reduced imports of goods compared to 2012 and 2011.

- The government continued to drag it back. Government at the state and federal level continued its contraction, subtracting 0.43 percentage points from growth as spending on everything from new purchases to national defense to teachers was scaled back for the third year in a row.

2014

- The United States economic recovery took a major step forward in 2014, achieving a number of important milestones.

- American businesses set a new record for the most consecutive months of job growth. By November, the economy had already added more jobs than in any full calendar since the 1990s.

- The pickup in job growth occurred primarily in higher paying industries, while nearly all the employment gains have been in full-time positions. At the same time, the unemployment rate fell below 6% for the first time since 2008.

2015

- The pronounced swing back toward urban living, and the renewed fervor for downtown development, made cities more dense making the search for cheap housing that much more challenging.

- A micro housing and micro dwelling boom hit big cities over the last few years, and while the number of developments remains relatively small, they've seen early success.

- Growing demographic and socio-economic change must be addressed to avoid becoming poorer and less competitive internationally.

- The continued strength in consumer sentiment was due to gains among households with incomes in the upper third of the distribution.
- The offshoring drive in manufacturing led to lower costs and access to new markets, but also created new problems for big companies when they ended up competing with their suppliers.

2016

- The West leads in the pace of job growth, but it also leads in terms of the types of jobs being created.
- Commodity prices continued to drag on the economies of the Oil Patch and Plains states.
- Home building and home sales remain one of the most important economic drivers across all of the country.
- The impact of soft global trade added to the weakness of the midsection of the country. The fall in demand and price for agricultural goods compounded the fall of energy commodities in the region, and much of its manufacturing is tied to domestic and global demand for agricultural equipment and supplies.

2017

- Wind power capacity in the United States continued to experience strong growth in 2017 due to the production tax credit, various state-level policies, and improvements in the cost and performance of wind power technologies yielding low-priced wind energy for utility, corporate, and other power purchasers.
- Solar panel technology delivers economy and environmental advantages in real-world conditions where it counts.
- United States' retail electricity sales fell by 80 billion kilowatt hours in 2017, the largest drop since the economic recession in 2009.

2018

- The Department of Commerce announced the availability of 587 million dollars for natural disaster aid.
- Ten projects with disaster related objectives were allotted a total of $85,980,000.
- The Department of Commerce invested in 113 projects, with non-disaster objectives, totaling $241,817,260. The objectives of those projects included...

- Manufacturing growth
- Entrepreneurship support
- Workforce training
- Infrastructure
- Business development
- Job growth
- Job creation
- Business growth
- Innovation growth
- Data processing center
- Flood protection
- Small business start-ups
- Aviation sector growth
- Bio manufacturing
- Industrial growth
- Technical education center
- Marine science and technology center
- Automotive training facility
- Manufacturing/Technology start-ups
- Economic growth
- Makerspace facility
- Boost tourism
- Shipping industry
- Renewable energy
- Ambulatory care center

2019

- The Department of Commerce announced the availability of 587 million dollars, the same amount as in 2018, for natural disaster aid.
- Seven projects with disaster related objectives were allotted $17,250,000.
- The Department of Commerce invested in 57 projects, with non-disaster objectives, totaling $167,952,000. The objectives for those projects included…

- Manufacturing growth
- Infrastructure
- Workforce development
- Technology development
- Entrepreneurship development
- Broadband expansion
- Flood protection
- Workforce training
- Industry growth
- Business growth
- Unmanned aviation
- Job loss
- Innovation center
- Cyber security
- Business loans
- Redevelopment

- The Department of Commerce invested in 53 projects with specific objectives, not related to natural disaster relief, without an indication of the dollar amounts allocated.

  At the end of 2019 the "experts" had conflicting views about economic development.

  - The economy will have a sluggish start in 2020, but will skirt a recession.
  - Better times are ahead, at least until 2022.
  - The economy is at a standstill, with no growth expected for 2020.
  - Data paints an unclear picture.
  - The economy heads into 2020 with steady growth.
  - Time will tell.

- Well, time has told! I am typing this last bullet on October 16, 2020. You can fill in the rest.

TECHNOLOGY DEVELOPMENT

- Technology is off and running! Some of the achievements and activities included...
  - Mobile Internet, Bluetooth and Wi-Fi
  - Great leaps were made in the area of robotic agility.

- Gene editing
- Smart watches
- Digitalization transformation in manufacturing for products, production systems, and product service.
- Tesla made the Model S vehicle into self-driving cars with a $2,500 upgrade that the cars downloaded over the air. This was one of the world's most significant software updates and in-app purchases.
- Space X unleashed the re-usable rocket!
- Tesla's Powerwall offered the ability with a high level of sophistication, letting you program your usage to collect energy during off-peak hours, and then consume it at peak-times.
- Supercomputers reached 100 petaflops.
- Microsoft launched the Windows 10 system.
- Electric car ownership reached one million worldwide, while trucks with emergency braking systems were made mandatory in Europe.
- HD CCTV cameras became ubiquitous and OLED displays came into widespread use.
- Agricultural robots became increasingly common on farms.
- The high bandwidth memory 2 standard was released by JEDEC.
- Fixstars Solutions released the world's first TB SSD.
- Scientists at MIT created the first five-atom quantum computer that had the potential to crack the security of traditional encryption schemes.
- Sales of electric and hybrid trucks reached 100,000 annually, while 10mm chips entered mass production.
- Electronic paper started seeing widespread use.
- Wireless, implantable devices that monitor health conditions in real-time came more into use.
- The enterprise-grade SSDs reached 100TB of capacity.
- Foldable phones and dual-screen laptops were on display.
- Next Fuel AB of Stockholm, Sweden, introduced the world's first Carbon Dioxide negative fuel at the UN Climate Summit in Katowice, Poland.
- Toshiba Memory America Inc. began sampling the industry's first Universal Flash Storage (UFS) Ver. 3.0 embedded flash memory devices.

- More than 300 million smartphones were shipped with some form of neural networking capabilities.
- 800,000 AI accelerators were shipped to data centers.
- Every day, 700 million people use some form of smart personal assistant, like an Amazon Echo or Apple's Siri.
- 60% of manufacturing plants today are focusing on technology investments.
- Lexar announced the first SD card that could store 1 terabyte.
- Uber, Lyft, and Airbnb were introduced.

EDUCATIONAL DEVELOPMENT

- The performance of America's students continues to be mediocre compared to other industrialized countries.

FUNDING OF EDUCATION

- Was still the responsibility of the states and localities.
- The United States spends significantly more on education than other OECD countries. In 2010, the United States spent 39% more per full-time student for elementary and secondary education than the average for other countries in the Organization for Economic Cooperation and Development (OECD).
- Yet, more money spent doesn't translate to better educational outcomes. In fact, American education is rife with problems, starting with the gaping differences between white students and students of color.
- Federal spending is limited to supporting special projects that attempt to address the mediocre performance of America's students (i.e., No child left behind, professional development for teachers and staff, benchmarks for student achievement, measurable goals for all students, etc.). The percent of the Total Federal Budget (Mandatory and Discretionary Spending) from 2010-2019 was...
  - 2010...3%
  - 2011...1.4%
  - 2012...2%
  - 2013...2%
  - 2014...3%

- 2015...3%
- 2016...2%
- 2017...2%
- 2018...2.6%
- 2019...2%

- That level of spending doesn't indicate a high priority item!

IMPORTANT EVENTS AND TRENDS

- Wars continued to drain financial resources and abuse the human resources of the United States.
- America's education system is not run by educators.
- American reform movements driven by millionaires and billionaires and free-market and privatization zealots with no real knowledge or understanding of public education continue to fall short of promised improvements (i.e., charter schools, cyber schools, vouchers, and corporate testing apparatus).
- Increasing pressure to integrate technology into the teaching process may reduce the role of teachers to that of an occasional proctor in the presence of students sent to a physical learning center (i.e., a computer lab).
- The interaction between student and teacher is where knowledge is developed. That knowledge is a major impact on any country's economic strength. That is why that interaction must be optimized, not diminished! If it is not optimized, America's education system will never be more than mediocre!
- America's population had grown from 308,745,538 in 2010 to 332,639,000 in 2020.

UNDERSTANDING THE PRESENT

2020
ECONOMIC DEVELOPMENT

- On October 23, 2020 the economy is in a tailspin as the Coronavirus spreads across America. The effects of the virus will be felt throughout the year and into 2021…or longer! The new normal is unknown…to be determined.

TECHNOLOGY DEVELOPMENT

- As 2019 came to an end, the "experts" identified the latest technology trends that would impact businesses in 2020.
  - 5G data networks
  - Autonomous Driving
  - Personalized and predictive medicine
  - Computer Vision
  - The Empowered Edge
  - The Democratization of Technology
  - Human Augmentation
  - Distributed Cloud
  - DARQ Age arrived – an asset for hiring and training
  - AI Products – for the ease of life
  - Data Policing
  - Momentary Markets
  - Automation for Advancements in Analytics
  - Medical Upgrades – the rise of 3D printing
  - Digital Debit
- Technology development is being hampered due to the mismatch between the skill sets of the Workforce Pipeline and the requirements of a Super Technology future.
- And then there was the Coronavirus!
- The focus is on developing a vaccine, possible treatments, bolstering the health care system's supply chains, and establishing rapid universal testing.

EDUCATIONAL DEVELOPMENT

- The education system is in shock! Home schooling via the internet is being implemented without proper preparation and validation.
- Hybrid systems, a combination of remote and classroom learning, are not working.

FUNDING OF EDUCATION

- Status quo…Primary responsibility is carried by the states and/or localities. The federal government budgets for special projects focused on specific attempts to resolve systemic problems.

IMPORTANT TRENDS AND EVENTS

- Wars continue to drain financial resources and abuse human resources of America.
- The Coronavirus is affecting every facet of America's culture.
- A conflict between state and federal leadership regarding managing the crisis has exposed critical problems within America's political systems and crisis management skills.
- After two decades of teaching teachers how to teach STEM, and trying to increase students' interest in STEM programs, teenagers' interest in STEM programs is decreasing.

CONCLUSIONS

- The federal government's indifferent attitude toward the Education System has created a situation where the skill sets of the Workforce Pipeline are not the skill sets required to be successful in a Super Technology economy.
- The STEM subjects are not being taught by teachers with extensive professional career experience in the subject they teach.
- Teenagers' interest in STEM careers is decreasing.

DESIGNING THE FUTURE

Designing STEM's future will be a mega project. The science of project management defines a process that consists of six major steps.

1. Project Statement…the project objective.
2. State the AS-IS…a description of the present situation.
3. State the TO-BE…a description of the ideal desired outcome.
4. GAP Analysis…identification of barriers that prevent the transition from AS-IS to TO-BE.
5. Create New Paradigms…to eliminate the GAP items.
6. Implementation…of new paradigms.

NEW PARADIGMS

Paradigm…A pattern, example, or model for success.

The title of this book…*The Future For STEM Is Grim*…determines the scope of the correction plan; fixing the STEM problem. The STEM problem is one of many problems that America's crippled republic must address. I have addressed those problems in a previous book…*Remaking America Is Up To You. The Future Is Outside The Box*, Copyright © 2019 by George Lopac, Jr…. with a number of "Out of the Box" new paradigms and alternate sources of funding to facilitate the implementation of the new paradigms. The alternate sources of funding can also be applied to the solution of the STEM problem.

Two new paradigms will establish a sound future for STEM.

1. Education Optimization
2. Synchronized Development of Economy & Workforce Skill Sets.

The success of each new paradigm depends on the success of the other paradigm. They are codependent.

EDUCATION OPTIMIZATION

If we are serious about optimizing education we must start at a very critical point in a child's life. In the first five years of life, experiences and relationships stimulate children's development; creating millions of connections in their brains. In fact, children's brains develop connections faster in the first five years than at any time in their lives. THIS IS THE TIME WHEN THE FOUNDATIONS FOR LEARNING, HEALTH AND BEHAVIOUR THROUGHOUT LIFE ARE LAID DOWN. This is the time when parenting of children must be done by parents in a family setting; not pre-school daycare!

PROJECT STATEMENT

Create a ready pool of workers (i.e., Workforce Pipeline) that are qualified and prepared to step up and fill relevant key roles within the economy as soon as they are identified.

AS-IS

A mediocre education system is incapable of providing the STEM skill sets required for success in a super technology global economy.

Teenagers' interest in STEM subjects/careers is decreasing.

TO-BE

The development of STEM skill sets within the Workforce Pipeline are in sync with the development of skill sets required for success in a super technology global economy.

Experienced professionals from the private sector teach STEM subjects; creating interest in STEM subjects/careers by sharing real-life experiences with the students.

GAP ANALYSIS

- A mediocre education system.
- Abuse of the education process.
- A Workforce Pipeline that does not contain the STEM skill sets required for success in a super technology global economy.
- The inability to tap into the skill sets of seasoned professionals in the private sector to become teachers of STEM subjects in the public education system.

CREATE NEW PARADIGMS

- A family will be paid an appropriate stipend per child, for the first five years of life, if parenting is provided by a stay-at-home parent. The amount of the stipend will be based on family structure and financial situation.
- Establish a Universal, free, Education System funded by the federal government (i.e., K-12, two-or four-year college, postgraduate, professional development, certification of specific skills, and adult education).
- The system is managed by educators, not business people, military leaders, or career politicians.
- Schools provide food, medical care, counseling, and transportation if needed.
- There is a national curriculum.
- All curricula are comprehensive and rigorous.
- Private schools must teach the national curriculum.

- Every school, public and private, has the same national goals and draws from the same pool of university trained educators.
- There are no classes for gifted students.
- The focus is on the student/teacher interface, where knowledge is developed.
- The mission is to maximize the development of each student's inherent potential and minimize the difference between the best student and the worst student.
- The maximum class size is ten students.
- There is no standardized testing.
- Diagnostic testing of students is used early and frequently. If a student is in need of extra help, intensive intervention is provided.
- There is no homework.
- Grades are not given until high school, and even then, class rankings are not compiled.
- Students are separated into academic and vocational tracks after the first two years of high school.
- Groups of teachers visit each other's classes to observe their colleagues at work.
- Teachers get one afternoon per week for professional development.
- Teaching is the most respected and most difficult degree to obtain.
- A master's degree in teaching is required to become a teacher in the system. Only the top ten percent of the graduating class will be considered.
- Teaching is the highest-status profession. Teaching careers have the highest salary levels of all careers. WITHOUT TEACHERS THERE WOULD BE NO OTHER CAREERS!
- TEACHERS MOLD AMERICA'S MOST VALUABLE RESOURCE, ITS PEOPLE, INTO A WORLD-CLASS EDUCATED WORKFORCE!

IMPLEMENTATION

Will depend on how well we participate in the democratic process. We must evaluate our representatives in congress; supporting those that share our values, replacing those that don't.

SYNCHRONIZED DEVELOPMENT
OF ECONOMY AND WORKFORCE SKILL SETS

PROJECT STATEMENT

Create a department whose sole responsibility is to ensure that the Education System is developing skill sets that are in sync with the skill sets needed for success in a future global economy driven by Super Technology.

AS-IS

No one is monitoring the skill sets being produced for the Workforce Pipeline, or the skill sets needed for success in a future global economy driven by Super Technology.

TO-BE

Skill sets being developed for the Workforce Pipeline are in sync with the skill sets needed for success in developing future economies.

GAP ANALYSIS

- A management blind spot.
- No awareness of the situation.
- No resources applied to correct the situation.

CREATE NEW PARADIGMS

- Create a Department of Skill Set Sync.
- The department will consist of five functions.
  - Project Management…Director
  - Education…Manager and staff
  - Business…Manager and staff
  - Technology…Manager and staff
  - Finance…Manager and staff
- Job requirements for Director and Managers
  - Minimum Age…fifty-five to sixty-five years
  - Term Limit…one five-year term

- ○ Professional Experience...thirty years minimum and recognition as an expert by his/her peers...cannot be an elected official or an employee in the public sector.
- Scope of activities
  - ○ Monitoring global technology development
  - ○ Monitoring global educational development
  - ○ Monitoring global economic development
  - ○ Monitoring United States educational development (i.e., Workforce Skill Sets)
  - ○ Quarterly status reports to Congress
  - ○ Incident reports as needed
  - ○ Recommended corrective actions as needed
- Selection of original department members.
  - ○ Congress will ask each of the subject areas in the private sector (i.e., Project Management, Education, Business, Technology and Finance) to recommend three candidates for each position. Recommendations cannot be elected government officials or regular government employees, Congress will select one candidate for each function.
- Replacement of department members.
  - ○ The department Director will ask each of the subject areas in the private sector (i.e., technology, education, economy, finance, and project management), to recommend three candidates for each of the open positions. Recommendations cannot be elected government officials or regular government employees. Department members will select one candidate for the open position/positions.
- All reports from the department to Congress will be available for public review.

IMPLEMENTATION

Will depend on how well we participate in the democratic process. We must evaluate our representatives in Congress; supporting those that share our values, replacing those that don't.

APPENDIX A
Author's Bio...George Lopac, Jr.

- Born and raised in Brooklyn, New York...1939
- Graduate of Brooklyn Technical High School's Aeronautical Engineering Program...1957
- BSME from the New York City College of Engineering...1962
- MSME from the University of Connecticut...1965

Thirty-Eight Years of Engineering Experience

- Experience with various global and domestic companies.
- Designed, developed, and manufactured retail and disposable medical devices.
- Designed, developed, and implemented manufacturing processes (i.e., fabrication, assembly, packaging, and inspection).
- Functioned at various levels of responsibility (i.e., Staff, Manager, Director, and Vice President).
- Conducted inter-company corporate training for major initiatives (i.e., Problem Solving, Project Planning and Control, Quality, and Common Sense).

Fourteen Years of Teaching Engineering

- Developed a unique system for maximizing student utilization of time.
- Developed a simple/effective system for Student/Peer evaluations.
- Contributed to the development and implementation of a rigorous, comprehensive STEM (Science, Technology, Engineering, and Mathematics) curriculum at a New Jersey high school; The Middlesex County Academy for Science, Mathematics, and Engineering Technologies. WE PUT THE E IN STEM!

APPENDIX B

The History of Brooklyn Technical High School
From the June 1954 Student's Handbook
A summary by Helen W. Codey

"It got started, this school of ours, in an idea, and all because a man dreamed a dream and saw a vision. He realized a need, had faith in a cause and persevered.

"Dr. Albert L. Colston, engineer, educator, saw the need of keeping pace in education with the intricate industrial system of the day. About 1916 he contemplated a change in the curriculum at Manual Training High School, where he was then chairman of the Mathematics Department. He was appointed chairman of a committee to investigate courses of study in technical high schools and to make recommendations concerning a new course at Manual.

"A technical course of study planned by Dr. Colston, was adopted by the Board of Superintendents in November 1918, and started at Manual Training High School in the spring of 1919, with Dr. Colston as director. At the end of the first year approximately 1200 pupils were enrolled in the course.

"With an increasing demand for this technical training, the idea of a separate technical school was seriously considered. In May 1922 the Board of Education decided to organize a technical high school on the basis of the course at Manual.

"In June Dr. Colston was nominated Principal of the new school. The loft building at Flatbush Avenue Extension and Concord Street was purchased and remodeled. There on September 11, 1922 the Brooklyn Technical High School opened its doors, with an enrollment of over 2400 pupils.

"At the start, football, basketball, cross country, and swimming teams were organized. The service squad (SOS), bank, Sales Bureau, and Survey staff were functioning. During that first term fifteen clubs were started, and school colors chosen.

"From the beginning the school building ran to capacity. Soon hundreds of applicants were being turned away. In November 1925 the Board of Superintendents recommended the acquisition of the present site for a new building. The site was approved for purchase January 1927; the appropriation for the building voted December 1928; plans approved May 1930;

and the contract awarded August 1930. In September 1930 ground was broken by Mayor Walker. Thirteen months later the cornerstone was laid.

"In September 1933 part of the new building was ready for occupancy, and about six hundred students began work there. As more rooms were equipped pupils were transferred from the old building and the annexes. Finally in June 1935 the old building was closed, and the fall term opened in September with the entire school a unit in the new building.

"This was the realization of Dr. Colston's dream. He had organized the courses, planned the building, selected the faculty, worked summers and winters, and persevered in the face of opposition, delays, and disappointments to bring about just this. The school stands a monument to a great organizer and administrator.

"Since 1935 Brooklyn Tech has grown in size, power, and reputation. The present enrollment is about six thousand. Extra-curricular activities; including clubs, teams, and service organizations number about 140. Some modifications have been made in the curriculum to keep pace with changing patterns in industry. Courses in aeronautics and industrial design have been added. The school's 10 KW radio station, WNYE, pioneered educational broadcasting in New York City for the Board of Education.

"During World War II a large number of our students and graduates qualified for the Navy College Training Program, Naval Air Corps, Army Specialized Training Program, and Navy Radio Technicians' Training Course. At the close of World War II an elaborate veteran's program was in progress.

"In June 1942 Dr. Colston retired. Ralph Breiling, administrative assistant, was appointed Acting Principal. In June 1945 William Pabst was appointed the second Principal of Brooklyn Technical High School.

"This is the Tech of yesterday and today. As for the Tech of tomorrow; may she grow in spirit, keep the faith, and see the job through."

A message from William Pabst, Principal of BTHS

"Brooklyn Technical High School has two special purposes. They are...

(1) To prepare students for advantageous entry into the technical branches of industry in the fields of aeronautics, architecture, industrial art, chemistry, electricity, and in the mechanical and structural fields.

(2) To provide a superior preparation for students who intend to enter engineering college.

"These purposes are accomplished through the provision of Seven Special Technical Courses and an Engineering College Preparatory Course. Certain basic principles are essential if a technical school such as Brooklyn Technical High School is to be successful. These principles are…

(1) Only high-caliber specially selected students capable of profiting by this more rigorous instruction are admitted to TECH.

(2) There must be no premature specialization. Students have the first two years in which to discover their special interests before choosing a course. During this period the broad foundation for all the courses is laid.

(3) There must be no blind-alley courses at Brooklyn Technical High School. The way to the top remains open! Any student of any course can go to college either by day or night, regardless of which course he has chosen.

(4) No theory or abstract principle is taught unless it is supplemented by appropriate illustrative shop or laboratory experience.

(5) There must be an especially selected and qualified faculty whose technical shop and drawing teachers have had appropriate practical experience as well as theoretical training.

(6) There must be up-to-date equipment, and present equipment must be constantly replaced by more modern equipment to keep up with the changes in technology.

(7) Last, though not least in importance, the technical student must have more than a scientific literacy. Competent, clear, concise, and interesting, spoken and written English, coupled with an understanding of our country's past and a study of present and future social and economic trends are imperatively needed assets of the young technician for effective life in a modern democracy. In addition, an appreciation of music and art will make him welcome in cosmopolitan social gatherings. Toward these ends, instruction in the aforementioned subjects is required in all courses.

"The success of Brooklyn Technical High School has been in direct proportion to its adherence to these principles. They represent the basic philosophy laid down by the founder of the school, Dr. Albert L. Colston, without whose vision and iron determination there would have been no TECH."

The TECH Creed...Class of June '24

"This is your school. Make it the best school in the city. Did you come to help or hinder? If you came to help, we want you. If you came to hinder, get out! Believe yourself a student in the best school in the city, in the best country in the world. A TECH man should be a hard fighter, a clean sportsman, a good winner, and a better loser. A TECH man should ask himself: 'What am I here for?' Now and forever afterwards he should remember that T-E-C-H stands for: T=truth; E=earnestness; C=courage; H=honesty."

Building and Facilities

The school, built on its present site from 1930-33 at a cost of $6 million, is twelve stories high, and covers over half a city block. Brooklyn Technical High School is directly across the street from Fort Green Park. Facilities at BTHS include:

- Gymnasia on the first and eighth floors, with a mezzanine running track above the first floor gym and a weight room on the third floor boy's locker room. The eighth floor gym had a bowling alley lane and an adjacent wire-mesh enclosed rooftop sometimes used for handball and for tennis practice.
- Twenty-five-yard swimming pool in the basement.
- Wood, machine, sheet metal and other specialized shops. A program involves a shop where an actual house is built and framed by students. Most have been converted into normal

classrooms or computer labs, except for a few robotics shops, such as the Ike Heller Computer Integrated Manufacturing and Robotics Center.

- A foundry on the seventh floor, with a floor of molding sand used for creating sand casting molds and equipped with furnaces, kilns, ovens and ancillary equipment for metal smelting. Students made wooden patterns in pattern making, which were used to make sand molds which were cast in the foundry and machined to specification in the machine shops. The foundry was closed in the late 1980s.

TECH's modernization would come under Principal William Pabst. Pabst created new majors and refined older ones, allowing students to select science and engineering preparatory majors including Aeronautical, Architecture, Chemical, Civil, Electrical (later including Electronics and Broadcast), Industrial Design, Mechanical, Structural, and Arts and Sciences. A general College Preparatory curriculum was added later. For the school year beginning in the last half of 1970, young women began attending TECH.

The Present

TECH is considered one of the country's most prestigious and selective high schools. With an enrollment of approximately 6,000, 1900 to 1950 students are admitted each year.

Current BTHS majors, taken from the school's website on July 28, 2020 are presented below. Majors are a set of selective courses given to students during their last twoyears at TECH, in a specific topic or discipline. These courses are in addition to those they would normally take to receive a New York City diploma.

Aerospace Engineering
Applied Mathematics
Architectural Engineering
Bio Science
Chemical Engineering
Civil Engineering
Digital Media
Electrical Engineering
Environmental Science

Finance

Industrial Design

Language Courses (French, Italian, Mandarin, Spanish)

Law and Society

LIU Advanced Health Professions

LIU PharmD

Mechanatronics and Robotics

Physics

Social Science Research

Software Engineering

Brooklyn Tech has achieved the rare feat of remaining true to its original goal of forming "a more technically literate workforce", while staying up-to-date with the latest technology and skills.

APPENDIX C
What Worked For Me
Overview

I started teaching at the Academy for Science, Mathematics, and Engineering Technologies in September of 2001. The Academy was a new school, and the first STEM high school in the Middlesex County Vocational and Technical School System. It is interesting to note that the high school I attended (i.e., Brooklyn Technical High School) was the first STEM high school in the country…no one knew it at the time since the acronym STEM was not created until 2001…and had its roots in the Manual Training High School in Brooklyn, New York.

The Academy's mission was to offer a rigorous, comprehensive, curriculum to develop the skills and knowledge that are prerequisites for success in an economy driven by technology. The engineering content of the program consisted of two engineering sections.

- Civil/Mechanical Engineering
- Computer/Electrical Engineering

I taught Civil/Mechanical Engineering, and another experienced engineer from the private sector taught Computer/Electrical Engineering. Since it was the first year of the Academy's history, there were no engineering curricula! We had to design curriculums from scratch, which is the best way for a teacher to utilize his/her experience.

I considered five issues before accepting the job at the Academy.

1. Administration…There was 100% support for the Academy project from the Superintendent and the Principal. They were relying on our professional experience to make the engineering components a benchmark for other schools…OK!
2. Facilities…A classroom approximately 24 feet by 36 feet…teacher's desk, chair, computer, printer, telephone, overhead projector, a large screen TV with a DVD/VHS player/recorder on a rolling cart, and a large white board. Seating for twenty students was provided with ten rectangular tables approximately 6 feet long and 2.5 feet wide, and twenty stand-alone chairs. That arrangement created the flexibility to replicate

manufacturing process flow and/or work stations in support of class projects. Two stand-alone steel cabinets (i.e., approximately 6 feet high x 3 feet wide x 2 feet deep), and a wall mounted cabinet, provided storage space for classroom supplies. A laboratory type sink provided access to hot and cold water. Each engineering section was allotted space in a teacher's storage room located on an upper floor. Each engineering classroom had two doors. One door connected to the main atrium, the other to a twenty-seat computer lab with a teacher's desk, computer, and a high capacity printer. We also had access to the Middlesex County College's gym; the Academy was located on the college's campus, for product, process, and experimental testing. Stocking an engineering program with appropriate tools and supplies would be up to me!...OK!

3. Schedule...Class met daily for ninety minutes (i.e., block schedule), the most efficient option...OK!

4. Curriculum...There were no curricula for Grades 9-12. The development of curricula would be up to me!...OK!

5. Salary...the salary gap between experienced engineers and experienced teachers is the major reason why STEM programs cannot get experienced engineers to teach engineering subjects in the STEM programs. There needs to be a way for experienced engineers to transition into teaching roles in a financially acceptable way. Fortunately I was able to deal with it!...OK!

The Academy was designed to handle forty students in each grade, 9-12. During year number one, the students were divided equally between the two engineering sections. At the mid-point of the year students switched engineering groups. At the end of the year students had to select a preferred engineering section to be their major for the next three years. During the first year I developed a curriculum for Grade 9 on the fly, and a curriculum for Grade 10. During the second year I developed a curriculum for Grade 11 while teaching Grades 9 & 10. During the third year I developed a curriculum for Grade 12 while teaching Grades 9,10, and 11. During the fourth year I taught all four grades, 9-12. Working with the students for three and a half years facilitated continuous development of the teacher/student interface, resulting in maximum development of the students' inherent potential.

The engineering sections ran on a block schedule, meeting daily for two consecutive forty-five-minute periods. You cannot get quality results meeting for only forty-five minutes at a time! The goal of each curriculum was to develop an understanding of why common sense is not common practice, how to be a team player, and to provide the opportunity to develop the skills and knowledge that are prerequisites for success in engineering.

Rigorous and Comprehensive Curriculum
Grade 9…Freshman Year

- Day #1 Design Challenge…(4) Teams of (5)…same starting materials
- Classroom Reading…*To Engineer Is Human*…Petroski.
- Structures…DVD…Article…students create ten quiz questions…quiz.
  - Skyscrapers
  - Bridges
  - Tunnels
  - Dams
  - Domes
- Materials…Article…Discussion…Students create ten quiz questions…quiz.
- Common Sense…Presentation…Discussion.
- Business Basics…Micro MBA…Presentation…Discussion.
- Industrial Processes…Presentation…Discussion.
- Quality… Assurance…Control…Presentation…Process Control Chart.
- Paradigms…DVD…Discussion.
- Synergy…Presentation…Discussion…2 + 2 = 5
- Engineering Drawing…Freehand
  - Isometric to Orthographic
  - Orthographic to Isometric
- Engineering Drawing…CAD
  - Freehand Isometric to CAD Isometric
- Engineering Disasters…DVD…Discussion
- Six Sigma…Presentation…Discussion
- Stress/Load/Area Relationship…Demonstration…Sample Problems…Problems to be solved

Grade 10...Sophomore Year

- Day #1 Design Challenge...(4) Teams of (5)...same starting materials
- Classroom Reading...*Lost Science*...Gerry Vassilatos.
- Research Assignments...to be worked on outside of class...Create a ten-to fifteen-minute presentation for a middle school audience...an introduction to the subject. Presentations will be made during the Class Activity period.
  - Bridge Engineering
  - Coastal & Harbor Engineering
  - Pumps & Compressors
  - Tunnel Engineering
  - Manufacturing Processes
  - Community & Regional Planning
- Class Assignments
  - Statics
  - Beams
  - Columns
  - Simple Machines
  - Electricity
  - Magnetism
  - Statistics
  - Thermodynamics
  - Hydrostatics
  - Truss Analysis
- Class Project...The class forms a company to Research-Design-Develop-Manufacture and document a product, product improvement, process, or process improvement of their choice.

Grade 11...Junior Year

- Day #1 Design Challenge...(4) Teams of (5)...same starting materials
- Classroom Reading...*Suppressed Inventions & Other Discoveries*...J. Eisen.
- Research Assignments...to be worked on outside of class...create a ten- to fifteen-minute presentation for a middle school audience...an introduction to the subject.

Presentations will be made during the Class Activity period.

- ○ Environmental Engineering
- ○ Highway Engineering
- ○ Jet Propulsion & Astronautics
- ○ Earth Work
- ○ Precision Measurements
- ○ Soil Mechanics
- ○ Aeronautics
- Class Assignments
  - ○ Kinematics (Cams)
  - ○ Programmable Logic Controllers (PLC)…Reference (Programmable Logic Controllers…Rabiee…Text, Lab Manual, and CD ROM)
  - ○ Springs
  - ○ Automated Manufacturing…3D Printing…Students print a simple component from a digital file.
  - ○ Programming With Alice… (Reference…Pausch…Text and CD ROM)…students create animated situations.
- Class Project…The class forms a company to Research-Design-Develop-Manufacture and document a product, product improvement, process, or process improvement of their choice.

Grade 12…Senior Year

- Day #1 Design Challenge…(4) teams of (5)…same starting materials
- Classroom Reading…*Nanotechnology for Dummies*…E. Boysen.
- Senior Project…Students work in teams of two to four members, each team will work on a different project, to Research-Design-Develop-Manufacture and document a product, product improvement, process, or process improvement of their choice. Each team must maintain a project log of all related activities. Each team will be responsible for…
  - ○ Quarterly Progress Reports
  - ○ Quarterly Design Reviews
  - ○ Quarterly Project Log Reviews

- Quarterly Seminars on a specific element of the project
- Train Project…Inventor CD ROM
  - CAD Assembly
  - Quarterly Progress Reports
  - Operation of the assembled train

Testing & Grading Formats Are the Same for All Grades

- Testing
  - Weekly Quizzes
  - Quarterly Tests…Any student achieving at least a 90% overall average for all Quiz grades has two options…OPTION 1: Be exempt from the Quarterly Test and get credit for (overall average for all Quizzes) X (5) Quarterly Test points. OPTION 2: Waive the exemption and take the Quarterly Test, hoping to get more points.
  - There is no Mid-Term Exam.
  - Final Exam…Any student achieving at least a 90% average on each Quarterly Test has two options. OPTION 1: Be exempt from the Final Exam and get credit for (overall average for all Quarter Tests) X (20) Final Exam points. OPTION2: Waive the exemption and take the Final Exam, hoping to get more points.
- Grading…The total year is worth 100 points. All grades are rounded to the nearest whole number…89.4 = 89 = B, and 89.5 = 90 = A.
  - 1st Quarter =20 points
  - 2nd Quarter = 20 points
  - 3rd Quarter = 20 points
  - 4th Quarter = 20 points
  - Final Exam = 20 points
  - There is no Mid-Term Exam.
- Within each Quarter
  - Homework = 2 points… (average of all Homework grades) X (2) Homework points.
  - Quizzes = 5 points… (average of all Quiz grades) X (5) Quiz points.
  - Quarterly Test = 5 points… (Quarterly Test grade) X (5) Quarterly Test points.
  - Participation/Performance…P/P = (8) points… (average of all P/P grades) X (8) P/P points.

○ Students are required to maintain a complete, and current, record of all grades received.

The Power of P/P Grades

The P/P grade is a powerful tool because it impacts 40% of each Quarterly grade, and Quarterly grades account for 80% of the final grade! It also provides students who struggle with tests a way to improve grades through focused participation and improved performance. For example: A classroom assignment was to solve ten problems with a two-week deadline. In this case each problem would be worth 10 points. However, there are a number of possible outcomes.

OUTCOME #1

I had nothing to hand in at the deadline…P/P grade = 0

OUTCOME #2

I submitted 10 completed problems on time…P/P grade = 100%

All 10 problems were correct…P/P grade = 100%

Documentation of my problem solving process was clear and neat…P/P grade = 100%

OUTCOME #3

I submitted 4 completed problems on time…P/P grade = 40 %

Three of the problems were correct…P/P grade = 30%

Documentation of my problem solving process was sloppy…P/P grade = 50%

The message is clear; if you do not participate and perform you will not succeed!

The Brain Buffet

The Brain Buffet was designed to develop the student's time management skills. It is a list of activities that students can work on for a given day. No student can say "I have nothing to do". The Brain Buffet was displayed on a magnetic white board in clear view of the class. At the top of the Brain Buffet was an 8 ½" X 11" piece of paper, landscape orientation, with a sketch of the human brain above the words; "Brain Buffet". Below that title sheet was the content of the Buffet written on the white board with a dry-erase marker. The content was updated for each grade on a daily basis during the Reading portion of the class.

General Content

Date…Changed each day

Reading…First 15 minutes of each class

Teacher's Presentation…15-30 minutes

Class Activities…45 minutes

- Date…Current Date

- Reading…An engineering related book selected by the teacher. The students were required to develop a process for distributing the books to the class, and returning the books into storage (i.e., a lesson in Process Development). That process starts as soon as they enter the classroom. If a student finishes reading the book, he/she may read another engineering related book of their choice that has been approved by the teacher.

- Teacher's Presentation…A video, physical demonstration, or discussion of the current engineering topic (i.e., Bridges, Stress and Strain, heat transfer, etc.) and distribution of a pertinent assignment. The demonstration could be how to solve a particular type of problem, and/or a physical demonstration of basic engineering principles. The distribution of an assignment could consist of a set of sample problems already solved, and a set of problems to be solved and submitted by a given due date.

- Class Activities…Students may work on any of the activities listed, as individuals, couples, or groups with no more than three members. It is not unusual for there to be multiple student arrangements working on multiple class activities at the same time. Typical Class Activities include…

  - Teacher's Assignment…A packet of problems to be solved, drawings to be made, models to be built, etc.

  - Major Design Project…For Grades 10 & 11 the class will act as a company, and organize into four departments…Research…Design/Development…Fabrication/Manufacturing…Project Management/Quality/Documentation. The company must develop a Project Plan and visually display and update the plan weekly. To facilitate visual display, an iron filled paint was used to convert part of a wall into a surface that would attract magnetized project planning elements.

  - Senior Project…Grade 12…Teams are formed and each team selects a project (i.e., new product, product improvement, new process, or process improvement). Each team must create a Project Book that documents the project events from start to finish. The Project Book must contain a Project Plan, experiments, data, drawings, all events, and weekly progress reports.

- Prep for Weekly Quiz
- Prep for Quarterly Exam
- Prep for Final Exam

Typical Brain Buffet Postings
Grade 9

Date…11/15/2010

Reading…*To Engineer is Human*…Petroski.

Presentation…Isometric Drawing

Activities

- Orthographic to Isometric Drawings
- Prep for Quiz

Grade 10

Date…11/15/2010

Reading…*Lost Science*…Gerry Vassilatos.

Presentation…Statics

Activities

- Research Assignment
- Statics Problems
- Design Project
- Prep for Quiz

Grade 11

Date…11/15/2010

Reading…*Automated Manufacturing*…J. Eisen.

Presentation…Cams

Activities

- Cam Drawings

- ○ Research Assignment
- ○ Design Project
- ○ Prep for Quiz

Grade 12

Date 11/15/2010

Reading...*Nanotechnology for Dummies*...E. Boysen.

Presentation...Design Reviews

Activities

- ○ Senior Project
- ○ Design Review
- ○ Project Book
- ○ Train Project...CAD
- ○ Prep for Quiz

The Peer Evaluation Process (PEP)

The PEP was conducted during Grade 10 at the end of the first marking period, and repeated at the end of the third marking period.

Instructions to the students...

This exercise will give you an idea of how your peers see you. It will be successful only if you are truthful and focus on behaviors, not specific situations that could be a clue as to your identity. The PEP is meant to be a "blind" exercise (i.e., no student's identity is known). Now is not the time to be a comedian or to be sarcastic. You can learn something about yourself and help somebody improve themselves.

The Process

- A class roster was given to each student on an 8 ½" X 11" sheet of paper.
- The names were listed vertically on the left side of the sheet.
- At each name there were four columns listed horizontally across the sheet: +...0...-... Negative Comments. Each student was required to check one of the first three columns.
  - ○ +=I always want this person on my team.
  - ○ -=I never want this person on my team.

- $\circ$ 0=I can work with this person.
- If the negative column is checked, the fourth column is used to indicate at least one reason to justify the rating. If more space is required, the flip side of the sheet was used.
- The students have twenty minutes to complete the evaluation.
- I collected the evaluations and prepared a summary sheet for each student that included the Histograms of the +, -, and 0 selections for all the students; without any student identification. On the flip side of that sheet were the negative remarks that were posted for the student receiving his/her summary sheet. Each student could compare his/her Histogram to those of the other students, and also see the reasons that justified the negative ratings.
- I met with each student, reviewed the results of the evaluation, and discussed what was required to reverse the negative comments.
- The PEP and the review process were repeated at the end of the third marking period.
- Most of the students that had received negative comments during the first evaluation were able to eliminate the root causes for the negative comments.

APPENDIX D

U.S. Patents Granted

Decade	Total U.S. Patents Granted
2010-2019	3,133,197
2000-2009	1,822,253
1990-1999	1,225,066
1980-1989	762,514
1970-1979	729,323
1960-1969	598,371
1950-1959	457,267
1940-1949	346,518
1930-1939	481,307
1920-1929	440,767
1910-1919	395,111
1900-1909	313,095
1890-1899	233,753
1880-1889	203,200
1870-1879	132,508
1860-1869	74,445
1850-1859	20,586
1840-1849	5,773
1830-1839	5,616
1820-1829	2,697
1810-1819	1,998
1800-1809	911
1790-1799	268

APPENDIX E

United States Wars

17th Century Wars

- Pequot War…1637
- King Phips War…1675
- Beaver Wars…Battle of Sorel…1610
- Anglo-French Conflicts…Battle of Port Royal…1613
- French and Iroquois Wars…1640-1701
- Battle of Fort Vercheres…1692
- Acadian Civil War…1643-1650
- Battle of Port Royal…1643
- Dutch Occupation of Acadia…1674
- Battle of Port La Tour…1677
- King William's War…1689-1697

18th Century Wars

- American Revolutionary War…1775-1783
- Cherokee-American Wars…1776-1795
- Northwest Indian War…1785-1793
- Shay's Rebellion…1786-1787
- Whiskey Rebellion…1791-1794
- Quasi-War…1796-1800
- Fries Rebellion…1799-1800

19th Century Wars

- First Barbary War…1801-1805
- German Coast Uprising…1811
- Tecumseh's War…1811
- War of 1812…1812-1815
- Creek War…1813-1814

- Second Barbary War…1815
- First Seminole War…1817-1818
- Texas-Indian Wars…1820-1875
- Arikara War…1823
- States…1825-1828
- Winnebago War…1827
- Black Hawk War…1832
- Texas Revolution…1835-1836
- Second Seminole War…1835-1842
- Second Sumatran Expedition…1838
- Aroostook War…1838
- Ivory Coast Expedition…1842
- Mexican-American War…1846-1848
- Cayuse War…1847-1855
- Apache Wars…1851-1900
- Bleeding Kansas…1854-1861
- Puget Sound War…1855-1856
- First Fiji Expedition…1855
- Rouge River Wars…1855-1856
- Third Seminole War…1855-1858
- Yakima War…1855-1858
- Second Opium War…1856-1859
- Utah War…1857-1858
- Navajo Wars…1858-1866
- Second Fiji Expedition…1859
- John Brown's Raid on Harper's Ferry…1859
- First and Second Cortina War…1859-1861
- Paiute War…1860
- American Civil War…1861-1865
- Yavapai Wars…1861-1875
- Dakota War of 1862…1862
- Colorado War…1863-1865

- Shimonoseki War...1863-1864
- Snake War...1864-1868
- Powder River War...1865
- Red Cloud's War...1866-1868
- Formosa Expedition...1867
- Comanche Campaign...1867-1875
- Korea Expedition...1871
- Modoc War...1872-1873
- Red River War...1874-1875
- Las Cuevas War...1875
- Great Sioux War of 1876...1876-1877
- Buffalo Hunter's War...1876-1877
- Nez Perce War...1877
- Bannock War...1878
- Cheyenne War...1878-1879
- Sheepeater Indian War...1879
- Victorio's War...1879-1881
- White River War...1879-1880
- Pine Ridge Campaign...1890-1891
- Garza Revolution...1891-1893
- Yaqui Wars...1896-1918
- Second Samoan Civil War...1898-1899
- Spanish-American War...1898
- Philippine-American War...1899-1902
- Moro Rebellion...1899-1913
- Boxer Rebellion...1899-1901

20th Century Wars

- Crazy Snake Rebellion...1909
- Border War...1910-1919
- Negro Rebellion...1912
- Occupation of Nicaragua...1912-1933

- Bluff War...1914-1915
- Occupation of Veracruz...1914
- Occupation of Haiti...1915-1934
- Occupation of the Dominican Republic...1916-1924
- World War I...1914-1918
- Russian Civil War...1918-1920
- Last Indian Uprising...1923
- World War II...1939-1945
- Korean War...1950-1953
- Laotian Civil War...1953-1975
- Lebanon Crisis...1958
- Bay of Pigs Invasion...1961
- Simba Rebellion on Operation Dragon Rouge...1964
- Vietnam War...1955-1975
- Communist Insurgency in Thailand...1965-1983
- Dominican Civil War...1965-1966
- Korean DMZ Conflict...1966-1969
- Insurgency in Bolivia...1966-1967
- Cambodian Civil War...1967-1975
- War in South Zaire...1978
- Gulf of Sidra Encounter...1981
- Multinational Intervention in Lebanon...1982-1984
- Invasion of Grenada...1983
- Action in the Gulf of Sidra...1986
- Bombing of Libya...1986
- Tanker War....1987-1988
- Tobruk Encounter...1989
- Invasion of Panama...1989-1990
- Gulf War...1990-1991
- Iraqi No-Fly Zone Enforcement Operations...1991-2003
- First U.S. Intervention in the Somali Civil War...1992-1995
- Bosnian War...1992-1995

- Intervention in Haiti…1994-1995
- Kosovo War…1998-1999
- Operation Infinite Reach…1998

21st Century Wars

- War in Afghanistan…2001-Present
- 2003 Invasion of Iraq…2003
- Iraq War…2003-2011
- War in North-West Pakistan…2004-Present
- Second U.S. Intervention in the Somali Civil War…2007-Present
- Operation Ocean Shield…2009-2016
- International Intervention in Libya…2011
- Operation Observant Compass…2011-2017
- American-Led Intervention in Iraq…2014-Present
- American-Led Intervention in Syria…2014-Present
- Yemeni Civil War…2015-Present
- American Intervention in Libya…2015-Present

APPENDIX F
U.S. Population Growth

Census Year	Population
2020	332,639,000…U.S. Census Projection from 2017
2010	308,745,538
2000	281,421,906
1990	248,709,873
1980	226,545,805
1970	203,211,926
1960	179,323,175
1950	151,325,798
1940	132,164,569
1930	123,202,624
1920	106,021,537
1910	92,228,496
1900	76,212,168
1890	62,979,766
1880	50,189,209
1870	38,558,371
1860	31,443,321
1850	23,191,876
1840	17,069,453
1830	12,866,020
1820	9,638,453
1810	7,239,881
1800	5,308,483
1790	3,929,214
1780	2,780,369
1770	2,148,076
1760	1,593,625

1750	1,170,760
1740	905,563
1730	629,445
1720	466,185
1710	331,711
1700	250,888
1690	210,372
1680	151,507
1670	111,935
1660	75,058
1650	50,368
1640	26,634
1630	4,646
1620	2,302
1610	350